Using Population Health Indicators for Global Health and Development

Raywat Deonandan

Intanjible Publishing

I've had the great fortune in my life to have traveled the world and to have met people in all social, political, geographic and economic circumstances. Many have become subjects of my research. But few, if any, have ever directly benefited from the fruits of that research. This book is dedicated to the crowds of nameless people who are the subject of population health research, but whose individual worth must never be denied or forgotten.

Contents

Chapter One

INTRODUCTION

This volume is an extension of two of my previous books, *Introduction to International Health Theory: An Interdisciplinary Perspective* [1] and *Nothing To Do With Skin: the Fundamentals of Epidemiology and Population Health Research.* [2] I've distilled and expanded upon elements from those books that many students found particularly useful, specifically around how population health issues are measured quantitatively. The astute reader will find substantial overlap between this volume and those earlier texts. This is an intentional attempt to focus on specific items from those books and to offer them in a lighter, less expensive package.

Policymakers and media increasingly blindly rely upon reported statistics both for shaping public opinion and for making official policy on important health matters. Yet often they remain unaware of the extent to which assumptions, agendas, and biases dictate the nature and content of those statistics. This is not to suggest that such statistics are developed in bad faith, but rather that their implementation requires a more nuanced understanding of their intent and foundation, such that they are not misused or misunderstood.

A good example is the oft heard mantra in Western countries, that we have an "ageing population." What exactly does this mean? Many feel that the phrase suggests that Western peoples are experiencing a dramatically longer lifespan. Others understand that it means that the average population age is higher than in previous generations, which further suggests that reproduction rates are declining. In other words, the phrase deftly avoids drawing a needed distinction between life expectancy and lifespan, which in turn can cause a skewed public understanding of the current demographic reality.

Measures of mortality and morbidity are similarly misunderstood. The distinction between incidence and prevalence, while largely inconsequential in everyday lay parlance, can have a dramatic impact on program evaluation and long term health planning. Similarly, competing indices of mortality ---proportionate mortality rate, case fatality rate (CFR), crude mortality rate, etc.--- each offers a particular perspective on the lethality of a given health condition, and none can individually convey the full thrust and import of that condition. Yet the scary estimates of a virulent disease's extremely high CFR can capture the public's imagination and profoundly overestimate that disease's importance. The single case of Ebola in the USA in 2014 is a fine example of this phenomenon.

While my larger goal is to enhance the knowledge and perspective of those engaged in population health research or policymaking, this volume is directed to a general lay audience with a very basic familiarity with health sciences terminology and issues. It is meant to offer a more rounded appreciation of some of the issues surrounding indicator development and usage. My intent is not to convey upon the reader expertise in population health measurement, but rather to empower the reader to begin to interrogate and question cited statistics in order to ascertain their limitations and underlying assumptions.

As always, I invite readers to inform me if any errors are found, or if there are any suggestions for any future editions. I can be reached via the publisher's website (intanjible.com)

With gratitude,

Raywat Deonandan, PhD

Ottawa, Canada

May, 2015

Chapter Two

INDICATORS OF POPULATION HEALTH

R esources are always limited, and opportunities for expending those resources are constantly multiplying. In the practice of population health, those resources might be medicines, health care workers, research dollars, or a public health education budget. The setting of priorities for how best to apply those resources is one of the major systemic applications of international health policy workers.

The philosophy behind this system of triage is variable. One can decide upon a strategy that minimizes mortality, or minimizes suffering, or minimizes economic impact. Or one can focus on a particular demographic, such as children, that one possibly deems to be a morally and economically more important target for investment. Often such decisions must consider the implications for international relations or security. For example, failure to spend money to stem a growing tide of climate change refugees may offend a neighbouring state that fears that they will have to absorb the costs of those refugees.

Sustainability is also a factor in this priority setting. Consider this entirely hypothetical scenario. A people might be hungry, so the temptation would be to spend limited money to buy food to feed the needy. But when that food is consumed, the underlying problem of hunger still persists. So perhaps it is more meaningful to spend that money on agricultural infrastructure investments, so that the population can increase its potential to produce food in the future. While this seems like a logical path, don't forget that people are hungry now, and that their suffering will not be abated by crops that will not be harvested till months hence. And if a decision-maker is subject to a short-term political cycle, meaning that his or her political power is dependent upon satisfying the short-term needs of voters, then there is a strong incentive to satisfy immediate needs and a disincentive against long-term planning.

Clearly, deciding on how best to apply limited resources to a population crisis, especially a health crisis, is a complicated ordeal for which there often isn't a universally accepted "correct" path. To help direct this process, we rely upon relevant, timely information. Often, the kind of information that population scientists depend upon comes in the form of indicators.

A useful definition of an indicator might be that it is a statistical value that provides an indication of the condition or direction over time of performance of a defined process or achievement of a defined outcome. Really, an indicator is just a number that "indicates" some real process or phenomenon in a population. And frankly, it doesn't even need to be a number. More about that when we look at qualitative indicators.

In my opinion, indicators in population health have three functions: To monitor the extent to which certain diseases or issues are affecting the population; to monitor the extent to which programs are succeeding; and to best inform decision making. A more cynical mind might be tempted to add a 4th function, to justify or rationalize a position that has already been decided upon. In other words, indicators can be misused for propagandistic purposes.

In our modern era of rapid data collection and processing, indicators of varying quality are widely available. The World Bank maintains a free library of international development indicators at its website (data.worldbank.org/indicator). These include indicators of agriculture and rural development (e.g., the percent of a nation's land that is arable), aid effectiveness (e.g., life expectancy at birth), climate change (e.g., annual amount of nitrous oxide emissions), economic policy and external debt (e.g., gross domestic product), education (e.g., percent of children out of school), and poverty (e.g., income share by lowest 20 percent).

Indeed, any measurable population characteristic can be considered an indicator. Deciding which indicators should be defined and collected is in itself a challenging task that can define or consume an entire career.

The United Nations maintains several repositories of country-level indicators. The Millennium Development Goals indicator have a dedicated website, [3] as do the UN's collection of social indicators [4]. The emerging Sustainable Development Goals have a specific set of indicators associated with them, [5] which will prove essential in the long term monitoring of interventions arising from these new goals. Individual UN agencies, such as UNICEF, UNAIDS, and the WHO maintain their own public databanks of indicators, relevant to their individual mandates, on their own websites.

One cannot, or should not, discuss population measurements without also taking time to consider te varying quality of data sources. Though a set of indicators may benefit from the imprimatur of a large organization, such as the WHO, association with an organizational name is no guarantee of data quality. Between nations, there is often heterogeneity in oversight, methodology, sampling strategy, and quality control. For example, the female literacy rate may be higher in one country than in another, but the gap may not be as severe as the numbers suggest. This is due to differing data quality. Did both countries apply a standard definition of literacy with equal vigour? Was the indicator computed from a representative sample of women in both populations? Was the determination of literacy applied in a similar way? For example, was it done via self-reports in one country and via direct testing in another? A host of forces, both known and unknown, act to introduce imprecision and error into our set of indicators.

The bottom line is that, while data of this nature adds much perspective and evidence to a decision-making process, they cannot be accepted as indicative of the fullness of truth in absence of a deeper consideration of data quality and context.

Chapter Three

CHOOSING AN INDICATOR

To appreciate the subsequent example, it is important to first understand two basic indicators of population mortality: the proportional mortality rate (PMR) and the case fatality rate (CFR).

A disease's PMR is the proportion of all deaths in a population that are attributed to that disease:

PMR = (The number of deaths from a given disease) / (Total number of deaths in that population)

It is usually expressed as a percentage and essentially tells us how much of a population's death burden is caused by a given illness or condition.

The CFR, on the other hand, is the proportion of people with a given disease who died of that disease:

CFR = (The number of deaths from a given disease) / (Total number of people who contracted that disease)

It, too, is expressed as a percent and tells us something of the severity of a disease *once it has been contracted*.

Now consider a hypothetical population that is beset by three fictional diseases: Geekulism, Nerd Fever, and Dorkitis. Their PMR and CFR mortality indicators are as follows:

Disease	PMR	CFR
Geekulism	8%	72%
Nerd Fever	2%	26%
Dorkitis	0.1%	55%

A decision maker, compelled to choose a disease to which to target his limited resources, and faced only with these data, would be forgiven for putting the bulk of his attention on Geekulism. Of the ones presented, it accounts for most of this society's deaths. And, if contracted, it has a 72 percent chance of killing its victim.

Now consider that there is a new, fourth disease, called Dweeb Syndrome. Its indicators are as follows:

Disease	PMR	CFR
Dweeb Syndrome	10%	100%

By these statistics, Dweeb Syndrome is much more concerning than either Geekitis, Nerd Fever, or Dorkitis. A decision maker would be well served to direct the lion's share of his resources to this new threat. However, upon closer inspection, we find out that Dweeb Syndrome is *100 percent fatal* to people over sixty-five years of age and only *1 percent fatal* to those sixty-five and younger.

When age groups are teased out of a rate, we call it an "age-specific" rate. In the case of Dweeb Syndrome, we have computed the age-specific CFR (for those younger and older than 65). How does the new age-specific rate assist the decision maker? In many societies, sixty-five is the mandatory retirement age. This means that Dweeb Syndrome, while horrifyingly lethal, is really only a threat to those who are not economically active in this population. The vulgar conclusion is that it is entirely defensible not to consider Dweeb Syndrome to be a priority concern, especially given this country's scarce resources, if it has no appreciable impact on the national economy.

The same logic can be applied if we drill down according to profession, education level, and geography, and even to assess rates specific to race or gender or disability status, as morally problematic as that sounds. We can therefore investigate specific demographic slices to either highlight or mask characteristics. For example, if we compare mortality rates of diseases just among patients of a certain age bracket, any age effects would be concealed.

Indicators are just numbers. When people use numbers, we inject our own values, agendas, biases, and ethics. The selection of specific demographic slices, or the decision to adjust indicator values to standardize for age or gender, for example, can be politically or value driven decisions, or they can be rational decisions made to elucidate more nuanced information from the numbers. The challenge is in remaining in the latter camp, while also being aware of the extent to which values and agendas taint that process.

The message here is that the usefulness of indicators is dependent upon the question being asked of them. Is the question "Which disease kills more of our citizens?" or is it "Which disease has the most serious impact on our economy?" Or indeed, is it "Which disease has the most serious impact on our demographic of greatest value?"

As well, many diseases do not kill, but instead render lingering illness and disability. For those, a mortality rate in no way captures their full human impact. Fortunately, we have other measures, most notably the Disability Adjusted Life Year (DALY), which seek to address that particular limitation. The dimensions of a population health challenge are multifaceted and nuanced. The selection of appropriate indicators to convey such nuance should be a well-considered task.

In essence, when choosing an indicator, one should ask, "What is it I am truly trying to measure?" The options include:

- How many people are dying?

- If not dying, how many people are "affected" (however I choose to define "affected")?

- How many people are suffering (however I choose to define "suffering")?

- What's it costing?

- What will it cost if I do not address it?

- How fast is it growing?

One useful way to categorize population health indicators is by slotting them five broad domains:

1. Health status, which provides information about the health of Canadians, including well-being, human function, and selected health conditions;

2. Non-medical determinants of health, which reflect factors outside of the health system that affect health;

3. Health system performance, for providing insight into the quality of health services, including accessibility, appropriateness, effectiveness, and patient safety;

4. Community and health system characteristics, that provide contextual information, not direct measures of health status

or quality of care; and

5. Equity, which is a cross-cutting dimension for the four above.

Here are some examples of indicators and into which domain they fall:

Indicator Domain	Example of an indicator
Health status	Prevalence rate of breast cancer
Non-medical determinants of health	Smoking rate
Health system performance	Wait times for knee replacement surgery
Community and health system characteristics	Population density
Equity	Potential rate reduction, e.g. the extent to which breast cancer rates would change if each socio-economic group in the jurisdiction experienced the rate of the most affluent socio-economic group.

Chapter Four

TRADITIONAL INDICATORS OF POPULATION HEALTH

W hile the number of indicators in common use is literally into the thousands, there are a handful of classical indicators that provide a foundation for assessing the general state of populations' health.

Mortality Rate

Mortality rate is the number of people who died, divided by the total number of people who were at risk of dying. It is usually given as a percent or as a rate per hundred thousand.

Example: In 2000, the malaria mortality rate in Djibouti was 119 people per hundred thousand.

An important caveat is that there are many different subtypes of mortality rates. We have already mentioned the PMR and CFR and also that we often like to be demographic-specific, such as when computing the age-specific mortality rate.

Two important demographic-specific indicators are the infant and child mortality rates. The infant mortality rate is the proportion of infants who die before their first birthday, while the child mortality rate is the proportion of children who die before their fifth birthday.

Two important adjectives to understand when wading into this topic are "neonatal" and "perinatal." There is some disagreement about the specific definitions of these two terms. But in general, "neonatal" has to do with newborns in the first twenty-eight days of life. "Perinatal" concerns the time around birth, usually five months before and one month after birth.

We focus on the neonatal and perinatal period for a number of functional and philosophical reasons. Along with maternal mortality rates, indicators of child mortality are often considered telling signs of a health system's robustness. I like to think that caring for its next generation is a priority for every human society. Therefore a failure to do so must be indicative of a profoundly struggling health care system.

Indeed, according to the WHO's World Health Report of 2010, the top five global killers of children are all preventable maladies:

1. Acute neonatal conditions, mainly preterm birth, birth asphyxia, and infections

2. Lower respiratory infections, mostly pneumonia

3. Diarrhea

4. Malaria

5. Measles

And about half of all those deaths tend to occur in the same five countries: India, Nigeria, Democratic Republic of the Congo, Pakistan, and China. [6]

Poverty Line

Contrary to popular belief, different countries employ different methods for measuring poverty. As a result, determining whether one population is "poorer" than another can be difficult. Many approaches are possible, including the determination of minimal nutritional standards, minimal living standards, income or purchasing potential, or internal comparisons to a population average. [7] Most commonly, a poverty line is determined by subtracting the costs of essential expenditures from the typical household income. In other words, the poverty line is often the minimum level of income deemed necessary to achieve an adequate standard of living.

The European Union employs a poverty line cut-off that is internally defined, meaning that poverty is determined relative to how well other people in your population are faring. Specifically, in the EU people falling below 60 percent of median income are said to be "at-risk-of poverty." [8] Canada also uses a relative measure to define poverty. Canadians rely on something called a "Low Income Cut-off" or LICO. [9] The approach is essentially to estimate an income threshold at which families are expected to spend twenty percentage points more than the average family on food, shelter, and clothing. According to one computation, in 2002 Canada's household LICO was $29,163 (after tax). [10] Families making under this amount would be considered "low income," or colloquially "impoverished." In general, however, LICOs vary with geography and family size.

Note that differing methods of measuring poverty can be separated into "relative" and "absolute" measures. A measure of relative poverty defines "poverty" as being below some internally defined threshold, as in the European system. Absolute poverty is a level of poverty as defined in terms of the minimal requirements necessary to afford minimal standards of food, clothing, health care, and shelter.

It is important to note that in a system defining poverty relatively, by definition, regardless of overall wealth of the society, there will always be someone living in poverty. So in a society of millionaires, the poorest millionaires would be deemed to be "impoverished." Absolute poverty, on the other hand, is functionally the absence of enough resources (money) to secure basic resources of life.

According to the 1995 World Summit on Social Development, [11] "absolute poverty" means "a condition characterised by severe deprivation of basic human needs, including food, safe drinking water, sanitation facilities, health, shelter, education and information. It depends not only on income but also on access to services."

Operationalizing an absolute definition is difficult. One UN method [12] recommends defining a family as being impoverished if they have an insufficiency of any two of the following:

1. Food: BMI must be above a certain threshold (usually sixteen).

2. Safe drinking water: Water must not come solely from rivers and ponds and must be available nearby (less than fifteen minutes' walk each way).

3. Sanitation facilities: Toilets must be near and accessible.

4. Health: Treatment must be received for serious illnesses and pregnancy.

5. Shelter: Homes must have fewer than four people living in each room. Floors must not be made of dirt, mud, or clay.

6. Education: Everyone attends school or learns to read.

7. Information: Everyone must have access to newspapers, radios, televisions, computers, or telephones at home.

8. Access to services: Vague access to education, health, legal, social, and financial services.

Prevalence

A measure of "morbidity," prevalence is defined as the percent of people who have a specific disease.

Example: In Panama in 1995, the prevalence of HIV was 0.9 percent of the adult population over fifteen years.

Incidence

Another measure of "morbidity," incidence is the rate of *new* infections of a given disease over a period of time. It can be given as a percent, but usually reported as a rate per thousand people.

Example: In Uganda, the incidence of HIV among adults fell from 7.6 per thousand in 1990 to 3.2 per thousand in 1998.

Maternal Mortality Ratio (MMR)

The MMR is also called the "obstetrical death rate" and is the proportion of women giving birth who die during or shortly after pregnancy.
It's usually given as number of deaths per hundred thousand live
births.

*Example: In 2000, there were approximately four hundred maternal
deaths per hundred thousand live births, worldwide.*

Like chid and infant mortality rates, the MMR is often used as
an overall measurement of a nation's health care system's effectiveness, mostly because the overwhelming majority of maternal deaths
are preventable with existing, mostly affordable, and well-understood
technologies, techniques, and interventions. It is telling that 99 percent of the world's maternal deaths occur in low-income countries.
[13] The worst rates of MMR are clustered in central and sub-Saharan
Africa, as per figure 1, taken from the United Nations Population
Fund (UNFPA) [14].

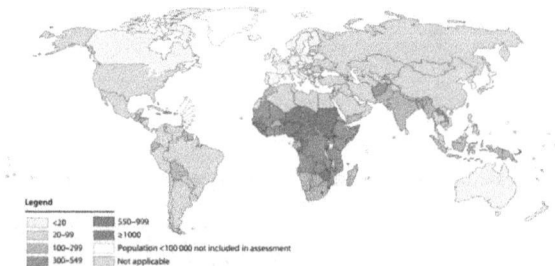

Figure 1. Map with countries by category according to their maternal mortality ratio
(MMR, death per 100 000 live births), 2010

Legend
<20   550–999
20–99   ≥1000
100–299   Population <100 000 not included in assessment
300–549   Not applicable

*Figure 1: Map with countries by category according to their
maternal mortality ratio (MMR, death per 100,000 live
births), 2010*

According to the WHO, [15] "in high-income countries, virtually all women have at least four antenatal care visits, are attended by a skilled health worker during childbirth and receive postpartum care. In low-income countries, just over a third of all pregnant women have the recommended four antenatal care visits."

Other factors that prevent women from receiving or seeking care during pregnancy and childbirth are:

- Poverty

- Distance

- Lack of information

- Inadequate services

- Cultural practices

Life Expectancy

The traditional "go to" indicator of a population's health is life expectancy. Even scientists who know of this measure's limitations will often resort to using it as a proxy of overall health, usually because life expectancies for many populations have been computed fairly rigorously for many decades now and therefore provide a ready method for comparing health trends over time.

According to the CIA World Factbook, based upon 2013 estimates, the ten countries with the highest life expectancies are:

Rank	Country	Life expectancy (years)
1	Monaco	89.63
2	Macau	84.46
3	Japan	84.19
4	Singapore	84.07
5	San Marino	83.12
6	Andorra	82.58
7	Guernsey	82.32
8	Switzerland	82.28
9	Hong Kong	82.20
10	Australia	81.98

Note that seven of those ten are essentially city-states or "micro nations," making their inclusion in the list questionable. Canada sits at #13 with 81.57 years, the United States at #51 with 78.62 years, and India at #164 with 67.8 years.

The bottom ten countries in their survey were as follows:

Rank	Country	Life expectancy (years)
214	Gabon	52.15
215	Namibia	52.03
216	Zambia	51.51
217	Somalia	51.19
218	Central African Republic	50.90
219	Afghanistan	50.11
220	Swaziland	50.01
221	Guinea-Bissau	49.50
222	South Africa	49.48

The bottom ten countries have some obvious things in common. All but one is a sub-Saharan African nation that has been brutalized by poverty, insecurity, and HIV/AIDS. And the exception, Afghanistan, has been war torn for decades now. But while those struggling nations are worthy of our concern, evidence suggests that globally life expectancy is on the rise, with men living eleven years longer, and women twelve years longer, than they would have forty years ago. [16]

There are three important considerations to keep in mind when considering how life expectancies were calculated:

1. The indicator does not distinguish between manners of death. Whether the population is dying from natural causes, disease, traffic accidents, war, crime, or natural disasters, the average life expectancy diminishes with every death.

2. Life expectancies are computed for a hypothetical, synthetic population. Therefore a population's life expectancy reflects the present state of affairs and is not meant to be a prediction of how long people will actually live.

3. The indicator represents the average experience of the previous age cohort. This means that experiences earlier in life tend to have a greater effect on total expectancy than experiences later in life. For example, infants *not dying* in early life will add a considerable amount of person-years to the total life expectancy, when computed at a point mid-life, as compared to the elderly, who will only add a few person-years.

As per point #3 above, dramatic improvements in preventing infant and child mortality are largely responsible for the observed improvements in life expectancy in many nations in the twentieth century.

Consider figure 2, which was taken from Robert Freitas, 2004. [17] It shows US life expectancies from 1850 to 2000, computed for three points in life: birth, age forty, and age seventy.

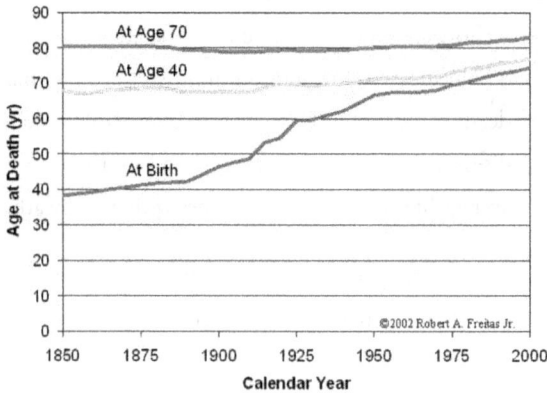

Figure 2: Life expectancy in the United States from 1850 to 2000

Traditional catalogues of life expectancy are given for computations at birth. By this standard, the improvements from 1850 to 2000 have been astounding indeed, with the American population gaining over forty years of lifespan. However, when computed in middle age, it is clear that gains in life expectancy over the past 150 years have actually been quite minimal. In other words, for most of human history, and in most struggling nations, if you're going to die, you are likely to die very young.

Early investments in the American public health infrastructure—clean water, hygienic toilet facilities, neonatal care, etc.—resulted in dramatically decreased infant and child mortality rates, further resulting in a significant improvement in life expectancy calculated at birth.

Therefore, when we say "our population is ageing," we don't mean so much that human longevity is increasing, but that there are more old people present, as a proportion of the total population.

Many low and middle income countries (LMICs) are presently undergoing this same shift, and we will revisit this phenomenon when we discuss the demographic transition. Consider figure 3 (which was derived from the World Bank Indicators, 2003) as confirmation of this global trend, as it shows that the emerging economic engines of the world, which were all very recently LMICs, are experiencing dramatic improvements in life expectancy.

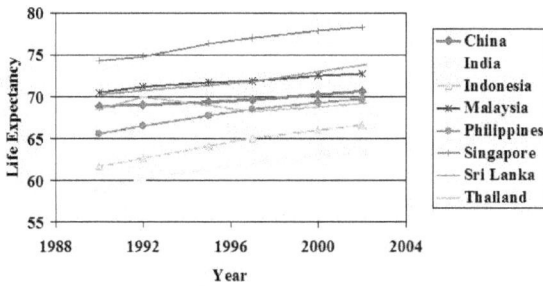

Figure 3: Life expectancy (calculated at birth) for selected Asian countries, 1988 to 2004

There are at least two profound problems when using life expectancy as an indicator of overall societal health:

1. A longer life does not necessarily mean a physically healthy or happy life

2. A longer life does not necessarily mean an economically productive life

Again, we must consider the underlying question that drove us to seek out an indicator in the first place. What is best for a population may not be what is best for an individual, depending upon the stark priorities of a decision maker.

Leaving aside the economic consideration, let us consider the need to incorporate quality of life into an indicator of life expectancy. In 1997, WHO director general Dr. Hiroshi Nakajima said, "Increased longevity without quality of life is an empty prize. Health expectancy is more important than life expectancy." With this in mind, some theorists have proposed a variety of modifications of the life expectancy indicator. One such variation is the "Health-Adjusted Life Expectancy" (HALE). As defined by the Australian Institute of Health and Welfare, the HALE is "an estimate of the number of healthy years (free from disability or disease) that a person born in a particular year can expect to live, based on current trends in deaths and disease patterns. The average number of years spent in unhealthy states is subtracted from the overall life expectancy, taking into account the relative severity of such states."

The HALE is closely related to the Healthy Life Years (HLY) indicator, which was developed by the European Union. Computed by Eurostat, "It is based on limitations in daily activities and therefore measures the number of remaining years that a person of a particular age can expect to live without disability." [18]

An average sixty-five-year-old man in Greece in 2000 would have expected to live an additional 16.1 years. His total life expectancy (TLE) would have therefore been 81.6 years. But only 9.5 of those additional years (HLY=9.5) were expected to be in good health. [18] Therefore the Greek HALE, computed at age sixty-five, in 2000 would have been 65+9.5 = 74.5 years. That's a difference of 7.1 years.

This is useful in at least two ways. First, it allows us to compare countries in a more useful way. An average sixty-five-year-old British man in 2000 would have expected to live an additional 15.8 years, giving a TLE of 80.8 years. Only 11.5 of those years, post-sixty-five, were HLYs. [18] Therefore, his HALE computed that year was 76.5 years.

Looking just at TLEs, one might conclude that Greek men were living 0.8 years longer than British men. But the proportion of HLY/TLE for the Greek men was 59 percent (9.5/16.1), whereas it was 73 percent (11.5/15.8) in the UK. So Greek men might have been slightly outliving British men, but British men were spending their later years in better health than were Greek men.

This is also useful in tracking changes within the same population. In the UK in 2000, the ratio of HLY/TLE for men was 73 percent. The next year it was 71 percent. By 2005, it was 61 percent. [18] Clearly there is a downward trend over time in Britain's recent history, suggesting that elderly UK men are spending fewer years in good health.

QALYs

Our next two indicators are conceptually similar. DALYs and QALYs are technically similar in that they both express health in time and give a weight to years lived with a disease, ultimately reflecting the same philosophy as the HLY and HALE above. It's important to remember that DALYs measure health loss and QALYs health gain. In other words, DALYs are bad and QALYs are good.

Invented in 1956 by economists Christopher Cundell and Carlos McCartney, QALY stands for "Quality Adjusted Life Year" and essentially measures how much "quality time" an intervention or event granted an individual or population. The QALY is intended to measure both the quality and the quantity of years granted by such an intervention. The QALY was originally developed as a measure of health effectiveness for cost-effectiveness analysis. [19]

In practice, each year lived is weighted with a quality score from 0.0 to 1.0, wherein 1.0 is a year spent in perfect health and 0.0 is death. Establishing weights is done through a variety of methods, including the application of EuroQol's "EQ-5D" questionnaire [20] or through a visual analog scale in which respondents are asked to rate a state of ill health on a scale from zero to one hundred, with zero representing being dead and one hundred representing perfect health. The value of such a weight is technically called a "utility" by economists, hence the usefulness of QALYs for "cost utility" purposes.

Consider if a new heart valve saves a patient's life (i.e., he or she would have died without it) but diminishes quality of life due his or her new inability to play sports or run. Using a variety of possible techniques, a utility or weight of 0.7 is granted to this experience. Perhaps this patient lives another ten years—years he or she would not have otherwise had.

A traditional perspective would hold that the new valve granted him ten years of additional life. But how many quality years of life did it grant him?

Number of QALYs	= Years	X	Utility
	= 10	X	0.7
	= 7		

QALYs are typically used for cost-effectiveness analysis to inform health care decision making and to estimate the extra quantity and quality of life provided by an intervention, which can be useful in choosing between different interventions. Economists will typically calculate a "cost per QALY" of a given intervention so that a decision can be made about whether a given technology or program is worth its expense, based upon the quality of life it offers, and not upon whether or not it simply extends life.

The following table summarizes the advantages and disadvantages of using QALYs:

Advantages and Disadvantages of QALYs [21]

Advantages	Disadvantages
• Provides a framework for valuing the health gains associated with interventions	• Values assigned to the quality of life component of the QALY may not reflect the values of patients receiving the intervention
• Can be used to help guide priority setting	• Controversial—whose quality values should be used?
• Combines estimates of both the extra length of life gained from an intervention and the quality of the extra life gained	• May lack sensitivity within a disease area
• Allows comparisons of the effectiveness of one intervention for a problem with the effectiveness of another intervention for the same problem	• Can oversimplify complex health care issues and suggest "quick and easy" resource allocation decisions
• Allows comparisons across disease areas to help show which programmes provide the greatest allocative efficiency	

DALYs

Like the QALY, the DALY—or "Disability Adjusted Life Year"—represents a conceptual and functional revolution in population health research. Prior to its innovation, the impact of a given detrimental event—a disease or terrible event—would most often and easily be expressed and measured as a mortality outcome. Thus, diseases that kill would always appear more impactful than those that maimed but did not kill.

The formula for computing a DALY is straightforward: $DALY = YLL + YLD$

Where YLL equals years of life lost in a population due to death and YLD equals years of productive life lost due to disability rather than death. Clearly, the computation YLL relies upon a satisfactory measure of expected lifespan; while YLD, much like in the case of the QALY, depends upon an estimation of the impact of a disease or event on quality and productivity of life.

The DALY has emerged to be the preferred indicator for comparing the impact of various diseases at the level of large populations. Its innovation allows us for the first time to compare the relative impacts of previously incomparable phenomena, such as bubonic plague versus schizophrenia; the first kills more people, but the second affects the quality of life for more people. Malaria is a good example of a disease whose effects can be better understood through a DALY lens. Varieties of malaria are quite fatal, but its most prevalent forms tends not to kill; those forms, though, render their victims disabled and economically unproductive.

In particular, the DALY has allowed mental illness to take its place amongst the panoply of global health concerns. The following table uses 2004 WHO surveillance data to compare the relative rankings of serious health threats according to both mortality and DALYs. The most glaring difference between the two lists is that depression, which does not appear to be a global mortality risk, is nevertheless responsible for the third highest share of DALYs in the world.

Leading causes of mortality globally (mortality rates), 2004		Leading causes of disease burden (DALYs), 2004	
1. Ischaemic heart disease	12.2%	1. Lower respiratory infections	6.2%
2. Cerebrovascular disease	9.7%	2. Diarrhoeal diseases	4.8%
3. Lower respiratory infections	7.1%	3. Depression	4.3%
4. COPD	5.1%	4. Ischaemic heart disease	4.1%
5. Diarrhoeal diseases	3.7%	5. HIV/AIDS	3.8%
6. HIV/AIDS	3.5%	6. Cerebrovascular disease	3.1%
7. Tuberculosis	2.5%	7. Prematurity, low birth weight	2.9%
8. Trachea, bronchus, lung cancers	2.3%	8. Birth asphyxia, birth trauma	2.7%
9. Road traffic accidents	2.2%	9. Road traffic accidents	2.7%
10. Prematurity, low birth weight	2.0%	10. Neonatal infections and other	2.7%

Data source: WHO health indicators, 2004 [22]

Given the DALY's sensitivity to societal impacts beyond mortality, it is a useful indicator as the outcome when modelling emerging global health concerns in coming years and decades. Several studies have attempted to predict the set of most impactful health concerns in our near future, keeping in mind current trends and expected shifts in demography, economics, and environment. One such study projects a shift in concerning health issues away from our current global concern with largely infectious agents to one focused more on chronic disease, as per figure 4, which was taken from WHO Health Indicators, 2004 [22]. This reflects a phenomenon called the epidemiologic transition, which will be discussed later.

2004 Disease or injury	As % of total DALYs	Rank	Rank	As % of total DALYs	2030 Disease or injury
Lower respiratory infections	6.2	1	1	6.2	Unipolar depressive disorders
Diarrhoeal diseases	4.8	2	2	5.5	Ischaemic heart disease
Unipolar depressive disorders	4.3	3	3	4.9	Road traffic accidents
Ischaemic heart disease	4.1	4	4	4.3	Cerebrovascular disease
HIV/AIDS	3.8	5	5	3.8	COPD
Cerebrovascular disease	3.1	6	6	3.2	Lower respiratory infections
Prematurity and low birth weight	2.9	7	7	2.9	Hearing loss, adult onset
Birth asphyxia and birth trauma	2.7	8	8	2.7	Refractive errors
Road traffic accidents	2.7	9	9	2.5	HIV/AIDS
Neonatal infections and other*	2.7	10	10	2.3	Diabetes mellitus
COPD	2.0	13	11	1.9	Neonatal infections and other*
Refractive errors	1.8	14	12	1.9	Prematurity and low birth weight
Hearing loss, adult onset	1.8	15	15	1.9	Birth asphyxia and birth trauma
Diabetes mellitus	1.3	19	18	1.6	Diarrhoeal diseases

Figure 4: Ten leading causes of burden of disease, world, 2004 and projected for 2030

Chapter Five

OVERPOPULATION

Its Negative Consequences

When discussing the plight of the developing world with non-specialists, one topic never fails to make itself known: the sheer mass of humanity on the planet. To many, the very complement of people is at the heart of most, if not all, human social ills.

For an area to be considered overpopulated, its population reaches a point where it can't be maintained without rapidly depleting non-renewable resources. In short, if its current human occupants are clearly degrading the long-term carrying capacity of an area, then that area is overpopulated. [23]

Overpopulation brings with it a host of potentially negative outcomes. They include:

- Ecological degradation

- Land overuse

- Diminished food and water supply

- Greater economic demands

- Increased population density means easier spread of communicable disease

- Potential for mass migration (i.e., refugees), which are a drain on the destination community

- Potential for border insecurity, usually caused by either mass migration or illegal immigration

Population Density

"Population density" is a measure of the concentration of people per a defined geographical unit. In truth, there are several measures of population density. For example, if one were to employ the entire surface area of the Earth then there are about thirteen people for every square kilometre. But of course, most of the surface of our planet is water. So if we only look at land masses then there are about forty-eight people per square kilometre. And if we exclude the continent of Antarctica, where no one but a handful of scientists reside, the estimate rises to fifty-three people were square kilometre.

Both estimates are probably not representative of most people's lived experience. Clearly, as an index, population density is flawed. Not all land is livable, and people tend to cluster in dense settlements. Indeed, it is estimated that an important milestone in human social evolution occurred sometime around 2008, when for the first time a majority of human beings lived in cities. [24] Predictions include that 64.1 percent of the developing world, and 85.9 percent of the developed world, will be city-based by 2050. [25]

Human distribution is too complicated to be summarized by a single measure. Nevertheless, we continue to attempt to do so. Some common indices of population density include:

Measure of Population Density	Description
Arithmetic density	The total number of people / area of land measured in km^2
Physiological density	The total population / the amount of arable land
Agricultural density	The total rural population / amount of agricultural land
Residential density	The number of people living in an urban area / the area of residential land
Urban density	The number of people inhabiting an urban area / the total area of urban land
Ecological optimum	The density of population that can be supported by the area's natural resources

According to Wikipedia, which compiled its data from a variety of official sources, the most populated nations in the world are:

Nation	Population	Date of estimate
People's Republic of China	1,354,040,000	December 31, 2012
India	1,210,569,573	March 1, 2011
United States	316,149,000	June 30, 2013
Indonesia	237,641,326	May 1, 2010
Brazil	193,946,886	July 1, 2012
WORLD	7,095,000,000	June 30, 2013

Also from Wikipedia, the following are the most *densely* populated nations on Earth:

Nation or Dependent Territory	Population density (people/km²)	Date of estimate
Macau	20,069	December 31, 2012
Monaco	18,068	December 31, 2012
Singapore	7,546	July 1, 2012
Hong Kong	6,516	December 31, 2012
Gibraltar	4,250	2011
WORLD	48 (land only)	June 29, 2013

Where do some of the poorer countries rank in terms of population?

Rank	Nation	Population	Date of Estimate
142	Namibia	2,113,077	August 28, 2011
143	Lesotho	2,074,000	June 30, 2013
144	Slovenia	2,060,261	June 30, 2013
145	Macedonia	2,059,794	December 31, 2011
146	Botswana	2,024,904	August 22, 2011

There seems to be a correlation between population size and population wealth. More specifically, density is a strong predictor of economic health. This may not be a causal relationship, since wealthy centres of economic activity tend to attract immigrants looking for opportunities; that's how cities are formed. But there's no denying that poor centres have fewer people, and rich centres have more people.

It is important to note that in the history of human civilization, India and China have both usually had the top-rated economies in the world. It's only in the past two to three hundred years when those particular countries have struggled. Their emerging prominence on the world economic stage can therefore be seen as more of a return to lost prominence.

With GDP being associated with population, it is a dangerous game to propose dramatic population reduction as a development strategy, as it may result in dramatic wealth decline, as well. This is due, of course, to how we define wealth in our modern Keynesian economy, which in part springs from Karl Marx's observation that rising population can actually be an indicator and accelerator of wealth due to the role of human resources as a form of capital.

Malthus

No discussion of overpopulation is complete without mention of Thomas Robert Malthus, an eighteenth- and nineteenth-century economist most famous for his treatise titled, "Essay on the Principle of Population." It is from him that we get the term "Malthusian Collapse."

Malthus argued that food production, at best, grows in an arithmetic, or linear, progression. In other words, if we double the land dedicated to food production, we double the amount of food produced. On the other hand, he pointed out, population growth is a geometric function, what lay people refer to as exponential growth. Geometric growth is faster than arithmetic growth. Therefore, given sufficient time and absent any unpredictable factors, such as epidemics, wars, or natural disasters, a population will always outgrow its food supply.

Famine, Malthus believed, is inevitable. And with famine comes the rapid fall of civil society and the Malthusian Collapse of civilizations. As dire as this model might seem, some argue that it was the norm in the ancient world; though evidence is debatable, and it is impossible to factor out the extraneous effects of war, disaster, etc.

In response, a nineteenth-century philosopher, William Godwin, pointed out three truths that undermine Malthus's treatise:

1. There's plenty of land to farm.

In Godwin's time, there was no shortage of arable land for food production, so there was no practical ceiling on food production. Many would argue that modern agricultural techniques make this today's truth, as well.

2. Reproductive rates will not necessarily be constant.

In a perfectly fertile society without social effectors, population growth is theoretically geometric. But many factors, such as religion, wealth, illness, economic stability, and lifestyle conspire to make true growth more linear in many cases.

3. Due to attrition, population growth is not geometric.

Variable mortality rates, disasters, wars, and epidemics also contribute to the flattening of most populations' growth curves.

And, as noted earlier in this chapter, famed economist Karl Marx argued that labour is capital, implying that larger populations may be more economically wealthy, and therefore can afford to purchase food and other commodities.

In general, supporters of Malthus's argument feel that the land's inability to produce infinite amounts of food limits its ability to sustain large human populations; whereas, critics of Malthus believe in the potential of the free market to create wealth for everyone, and therefore in the ability of populations to purchase resources.

Chapter Six

THE DEMOGRAPHIC TRANSITION

What Is the Transition?

As we have seen in the examination of population densities around the world, there is an uneven distribution of peoples worldwide. How to account for the rise in population in some nations and not in others? Why are wealthy European nations not as populous as emerging economies in Asia? Consider figure 5, showing the changes in birth and death rates of the UK from 1700 to 1840, which was taken from The Peel Web. [26]

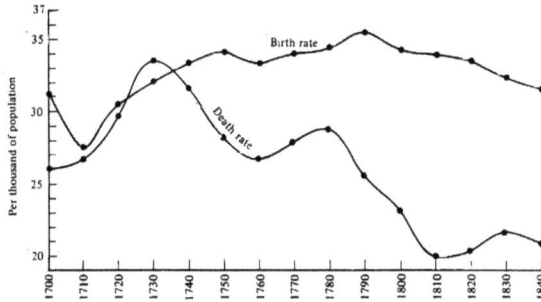

Figure 5: Changes in birth and death rate in UK, 1700 to 1840

Death rates rose slightly but remained relatively stable in the shown interval. But starting in the early 1700s, the death rate started to plunge and remained low from 1800 onward. The Industrial Revolution, investments in public infrastructure, hygiene, and food production slowed the death rate. As a result, the UK population boomed in that period.

But why then did the UK population not burst at the seams in the twentieth and twenty-first centuries? Their birth rate eventually declined, as well, leading to a somewhat stable population size.

This pattern was observed for several European countries. Consider figure 6, taken from Wikipedia, showing changes in Sweden's population from 1735 to 1995:

Figure 6: Demographic changes in Sweden from 1735 to 1995, CBR: Crude Birth Rate; CDR: Crude Death Rate

The pattern was so common that Warren Thompson developed a theory in 1929 that described a series of changes that he believed populations must undergo. In essence, he observed that a nation transitions from high birth rate and high death rate to low birth rate and low death rate as it "evolves" from a pre-industrial to post-industrial economy. He called this change the "demographic transition."

As labelled in the graph above of Swedish demographic change, the demographic transition occurs in four stages:

    1. Historically, pre-industrial society death rates and birth rates are high and roughly in balance.

    2. Death rates drop rapidly due to improvements in food supply and sanitation, which increase life spans and reduce disease. Total population therefore increases.

3. Birth rates fall due to a variety of social factors, but total population is still rising.

4. Both birth rates and death rates are low.

Some theorists provide for a fifth stage:

1. Total population is high but going into decline due to an ageing population.

Figure 7, taken from the British Broadcasting Corporation, [27] well summarizes the five stages:

Figure 7: The demographic transition's five stages

The Stages of the Demographic Transition

Stage 1

- In pre-industrial society, death and birth rates are high and fluctuate according to natural phenomena (drought, disaster, disease)

- Population is relatively young

- Cost of a child is the cost of feeding him or her (i.e., low)

- Economic contribution of child is high (working on farm, etc.)

- Net economic value of child is therefore high

- Majority of deaths concentrated in five- to ten-year-olds

Stage 2

- Decline in death rate, but birth rate remains high

- Increased survival of children, so age distribution shifts younger

- In Europe, initiated by Agricultural Revolution in eighteenth century

- Modern countries thought to be in stage 2: Yemen and Afghanistan, parts of sub-Saharan Africa (especially before the AIDS epidemic)

- Likely spurred by agricultural improvements, such as crop rotation and selective breeding

- Also spurred by public health improvements, such as vaccination, clean water, sewerage, maternal care

Stage 3

- Decline in birth rate begins

- Fewer children needed to maintain family economic unit due in part to mechanization of farming

- Increased urbanization

- Increased female literacy and education

- Improved contraception practices

Stage 4

- Called by some a "post-transition" stage

- Birth rates are more or less equal to death rates

- Some countries have a total fertility rate of <2.5

- Birth control is widely available and there is a desire for smaller families

Stage 5

- Also called "de-industrialization"

- Fertility rates are below replacement rate (i.e., national birth rate needed to maintain the total population)

The Demographic Transition in Non-European Nations

Adherents to the demographic transition model believe that countries presently experiencing population explosions will "transition"
to lower population states once appropriate economic, social, and
development advances are made or introduced. Figure 9, taken from
the Dementia Research Group, [28] shows the ageing trends of both
the UK and China:

*Figure 9: Actual and projected ageing trends in the UK
and China, 1890 to 2030*

China's age distribution is comparable to that of Canada and many
European nations, suggesting that it will reach stage 5 at roughly the
same time period.

The Demographic Trap

The so-called "demographic trap" occurs when a developing nation
does not reach a stable population or declining population levels
that are needed to enter the post-transition stage. Countries find
themselves "trapped," when there is a relatively low death rate and a
still-high birth rate.

The root of the problem lies in the resulting high natural increase of the population. The trapped country's economic growth ends up being used to support the needs of the booming population. There is little left over to promote the economic and social developments needed for further transition.

In other words, the country is stuck in stage 2—high birth rates and low death rates—resulting in a rapidly growing population. Perhaps Yemen is one such trapped nation. Stage 2 might persist because of falling living standards, which reinforce high fertility behaviours, which in turn reinforce the decline in living standards. This results in more poverty, and subsequently more reliance on having more children to provide economic security.

Consider figure 10, taken from the UN Food and Agriculture Organization, which shows overall explosive population growth in Yemen from 1965 to 2005:

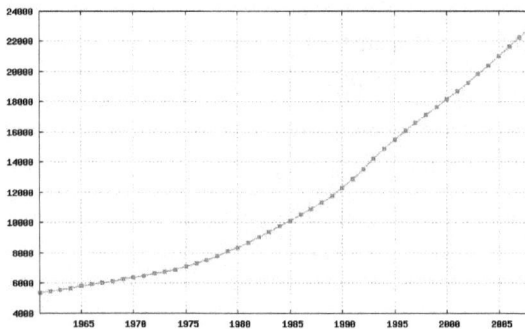

Figure 10: Population of Yemen from 1965 to 2005, in thousands of inhabitants

It is clear that Yemen's population is rising, but how much of this is due to fertility? To shed light on that question, we consider Yemen's fertility rate, shown in figure 11, compiled from the UN's Indicator Database, compared to that of neighbouring countries:

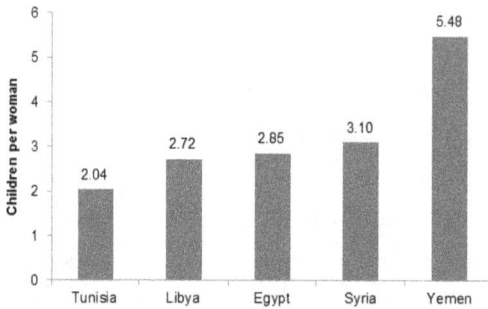

Figure 11: Female fertility rate of selected North African countries, 2012

According to current UN indicator data, 75 percent of Yemen's population is younger than thirty, compared to 60 percent in Egypt and 52 percent in Tunisia. The Yemeni population is a young one, with explosion happening at the lower ends of the age range.

Some theorists see the demographic trap as only a temporary problem that can be eliminated with better education and contraceptive behaviours. Others consider it more of a longer-term symptom of grander development failures, resulting in more families seeing children as economic investments. One thing all agree upon is that family planning must remain a crucial part of economic development plans.

Criticisms of the Demographic Transition Model

- The model was developed after studying the history of Europe. Cultural and historical considerations are different in other parts of the world, and the European model might not be valid for all circumstances. Countries like China, whose one-child policy dramatically and artificially altered that country's demographic profile, appear to have followed a short cut to an imminent stage 5. And other countries, whose reproductive practices are defined more profoundly by their cultural situations, which may affect such things as the acceptability of contraception and the rigid roles of women, may have intrinsic cultural resistance to the forces of the demographic transition.

- The transition model does not account for unforeseen game-changing developments, such as the advent of HIV/AIDS, which has disproportionately affected LMICs, and which has removed mostly young adults in their economic prime.

- Some economists argue that modern political forces prevent transitional forces from coming to bear. Subsistence farmers cannot transition to industrial economics, for example, because their local political structures prevent personal ownership of their land. And the modern reality of globalization, which effects international pressures on local manufacturing economies to service commodity needs of foreign post-industrial economies, may prevent the former from achieving their own post-industrial state.

Chapter Seven

EPIDEMIOLOGIC TRANSITION

I n 1971, Abdel Omran reasoned that the history of human mortality could be divided into three broad stages, as a population goes through the process of changing from an underdeveloped to a developed nation: [29]

1. Age of Pestilence and Famine

High mortality, therefore low population growth, low life expectancy

2. Age of Receding Pandemics

Declining mortality, fewer infectious epidemics, sustained population growth

3. Age of Degenerative and Manmade Diseases

Further declining mortality, chronic diseases become more important than infectious

He defined the shift through those ages as the "epidemiologic transition," which he described as "a human phase of development witnessed by a sudden and stark increase in population growth rates brought about by medical innovation in disease or sickness therapy and treatment, followed by a re-leveling of population growth from subsequent declines in procreation rates." [30]

Though historically, Omran's first age refers to pre-agricultural societies, one would argue that in today's world, many of the more struggling low-income countries would qualify. In some sub-Saharan African countries, for example, life expectancy is very low and infectious disease rates are high. In particular, infant mortality rates are high. Chad and Mali are good examples, each with infant mortality rates above 120 (deaths per thousand live births), based upon 2005 data.

The second age was historically characterized by advances in health care, particularly public health investments and breakthroughs, such as clean municipal water and the discovery of antibiotics. World population growth surged in the decades after the refinement of penicillin, with the world gaining two billion people between 1950 and the 1980s alone.

The third age occurs when birth rates decline rapidly to meet, and sometimes drop beneath, replacement rates. Omran identified [31] three possible factors that likely encouraged transition to the third age:

1. **Biophysiologic factors**

Reduced infant mortality and more people living into middle and old age

2. Socioeconomic factors

Economic value of smaller family sizes

3. Psychologic factors

Focus on quality of child-rearing, rather than on numbers of children and bare survival

A widespread criticism of Omran's model is that there might not have actually been a shift from infectious to chronic diseases (mostly in the twentieth century). Critics argue that detection bias, owing to new diagnostic techniques, changes in societal values, and more elderly people (and therefore more people likely to present with chronic illnesses) are the real causes of greater chronic illness, not a transition *per se.*

Chapter Eight

HUNGER

All theorists, Malthus included, relate population to food supply. The greatest threat of overpopulation is that a community will outgrow its ability to feed itself, leading to mass die-off. Of course, other effects, such as ecological degradation, disease, and war are also likely. But hunger is, in many ways, a pre-condition to those effects.

The concept of "food security" was therefore developed to describe a population's relationship with its food supply. The World Food Summit of 1996 defined it as "when all people at all times have access to sufficient, safe, nutritious food to maintain a healthy and active life." Commonly, the concept of food security is defined as including both physical and economic access to food that meets people's dietary needs as well as their food preferences. [32] Food security is built upon three pillars:

1. Food availability: sufficient quantities of food available on a consistent basis

2. Food access: having sufficient resources to obtain appropriate foods for a nutritious diet

Food use: appropriate use based on knowledge of basic nu-

3. trition and care as well as adequate water and sanitation

The following pie chart (figure 12), taken from World Hunger, [33] shows the geographical distribution of hungry people in the world. Note that an overwhelming majority are clustered in the Asia-Pacific region.

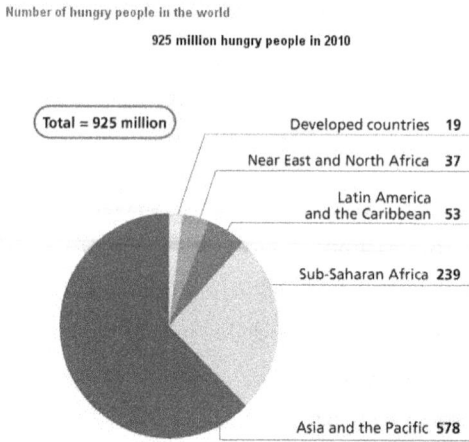

Number of hungry people in the world

925 million hungry people in 2010

Figure 12: Geographical distribution of the world's hungry, 2010

This is an important observation, given that the nineteenth and early twentieth centuries saw staggering famines in South Asia in particular; yet the twenty-first century has been characterized by the explosive economic growth of both China and India. And with greater wealth comes a greater demand for richer, high caloric, and less sustainably produced foods.

The International Food Policy Research Institute [34] projects that in the next twenty years, India alone will increase its demand for meat by 176 percent, milk and vegetables by 70 percent, and grain by 27 percent. This is due to the rapid expansion of that country's wealthy middle class, whose dietary habits increasingly resemble those of comparably wealthy Westerners.

The following table shows the change in the share of poor people in both India and China between 1990 (prior to those countries' economic emergence) and 2005 (well into the economic revival). Note that poverty in this case is defined as living on less than $1.25 per day, as per the World Bank's standard definition.

Percentage of the World's Poor Living in Asia

Nation	1990	2005
China	38%	15%
India	24%	32%
Rest of Asia	8%	43%

Source: "Renewed Policy Action for the Poorest and Hungry in South Asia," Dec 2, 2008, IFPRI [35]

While China's economic growth has resulted in a lessened share of the poor, India's has resulted in a higher share. It is unclear whether this is due to China's ability to leverage its new wealth to feed its poor or to India's more attractive role as an immigration destination for the region's poor. In any case, the lesson is that increased wealth alone does not seem able to quash a nation's burden of poverty, given regional crises and internal heterogeneities.

When deciding upon a nation's burden of hunger, a series of indicators are traditionally employed: agricultural production, the human development index, underweight children, rate of undernourishment, the so-called global hunger index, and the ever-important infant mortality rate.

The following graphic (figure 13), summarized from definitions given by the World Food Programme, summarizes the definitions of these indicators:

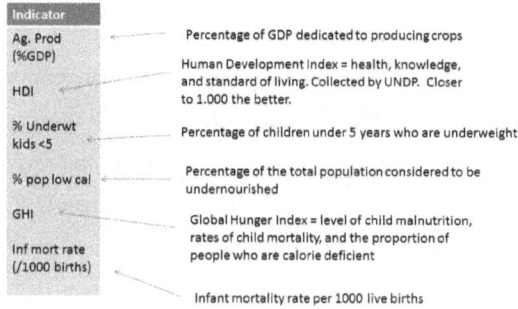

Figure 13: Hunger indicators

The HDI is a composite score that attempts to measure standard of living. The GHI is a composite score intended to provide a quick estimate of a population's struggles with hunger. If it is over thirty, the state of hunger is said to be "extremely alarming"; a value of twenty to thirty is said to be "alarming"; ten to twenty is merely "serious"; five to ten is "moderate"; and a value of less than five is basic "hunger."

The GHI is not traditionally calculated for very wealthy countries, like Canada and the United States. But a wealthy nation with a variable standard of living, like Saudi Arabia, would have a value somewhere between one and five.

A summary of the above health indicators applied to Canada would yield the following values (figure 14):

Indicator	CANADA
Ag. Prod (%GDP)	2
HDI	0.967
% Underwt kids <5	<1
% pop low cal	<1
GHI	n/a
Inf mort rate (/1000 births)	5

Figure 14: Canada's hunger indicators

Meanwhile, selected South Asian nations offer the following values (data extracted from the World Food Programme in 2010) [36] summarized in figure 15:

Indicator	Bangladesh	India	Pakistan	Sri Lanka	Nepal
Ag. Prod (%GDP)	32%	60	20	11.7	43
HDI	0.547	0.619	0.551	0.743	0.534
% Underwt kids <5	48%	47	38	29	48
% pop low cal	30%	20	24	22	17
GHI	25.2	23.7	21.7	15	20.6
Inf mort rate (/1000 births)	54	56	79	12	56

Figure 15: Hunger indicators in South Asia

What leaps out from these statistics is that Sri Lanka has a traditionally high HDI, even compared to its economic giant of a neighbour, India. Sri Lanka does well in other measures, as well, but has been hobbled in recent years due to its civil war.

India, meanwhile, despite being a regional military and economic superpower, is struggling on many development fronts, most notably in child hunger and mortality. So, while its middle class is posting remarkable wealth statistics, its children and lower class are nevertheless languishing in much lesser development states.

The relationship between population, hunger, and wealth is not a simple one. The Malthusian view holds that food production is the gateway to feeding a growing population. Economists like Nobel Prize-winner Amartya Sen argue that food existence is irrelevant if food access is not well established; wealth is one gateway to access. But the composite indicators of population hunger also suggest that overall population wealth is insufficient as a predictor of general hunger.

Chapter Nine

QUALITATIVE INDICATORS

N umeric indicators are without question the most commonly collected and expressed, at least to the extent that they are represented in government reports and databases. But it is important to acknowledge that qualitative indicators have an important role to play, as well.

Qualitative indicators are, essentially, individuals' judgments or perceptions about a subject. For example, "the increase in the level of confidence subjects have in their ability get timely health care, or "how much did fear of violence change in the past year."

Very often, qualitative indicators have a numerical component, as well, and are often merely a quantitative expression of a qualitative sentiment. The examples above can be made to include a numeric component by considering the *percentage* of people who feel they can get timely health care, or the *proportion* of people who fear violence less.

Chapter Ten

A FINAL WORD

T his is, of course, not an exhaustive list or investigation of the indicators used in population health research. There are ongoing efforts to define and refine health indicators, as well as to develop new ones. Current activity in this area concerns measurements of mental health, fertility, and out-of-pocket expenditures on health care. For further reading on this topic, I recommend browsing the Center for Global Development's working group report on global heath indicators. [37]

As noted, I welcome all comments, complaints and suggestions regarding the content of this book. You may reach me via the publisher's website.

Chapter Eleven

REFERENCES

[1] R. Deonandan, Introduction to International Health Theory: An Interdisciplinary Perspective, Kendall Hunt, 2013.

[2] R. Deonandan, Nothing To Do With Skin: the Fundamentals of Epidemiology and Population Health Research, Deonandan Consulting Inc, 2014.

[3] United Nations Statistics Division, "Millennium Development Goals Indicators," 2014. [Online]. Available: http://mdgs.un.org/unsd/mdg/Default.aspx.

[4] United Nations, "Social Indicators," [Online]. Available: http://unstats.un.org/unsd/demographic/products/socind/.

[5] Sustainable Development Solutions Network, "Indicators and a Monitoring Framework for Sustainable Development Goals: Launching a data revolution for the SDGs," 2015. [Online]. Available: http://unsdsn.org/resources/publications/indicators/.

[6] B. e. al, "Global, regional, and national causes of child mortality in 2008: a systematic analysis," The Lancet, vol. 375, no. 9730, pp. 1969-1987, 2010.

[7] M. Ravallion, "Poverty Lines across the World," The World Bank Development Research Office, Washington, DC, 2010.

[8] "The European Anti-Poverty Network," [Online]. Available: http://www.eapn.eu/en/what-is-poverty/how-is-poverty-measured. [Accessed 7 june 2013].

[9] S. Canada, "Low income cut-offs," [Online]. Available: http://www.statcan.gc.ca/pub/75f0002m/2009002/s2-eng.htm. [Accessed 7 june 2013].

[10] G. deGroot-Maggetti, "A measure of poverty in Canada," Citizens for Public Justice, 2002.

[11] United Nations, "World Summit for Social Development," 1995. [Online]. Available: http://www.un.org/esa/socdev/wssd/text-version/.

[12] D. Gordon, "Indicators of Poverty & Hunger," 2005. [Online]. Available: http://www.un.org/esa/socdev/unyin/documents/ydiDavidGordon_poverty.pdf.

[13] "Trends in Maternal Mortality: 1990 to 2008," World Health Organization, Geneva, 2010.

[14] United Nations Population Fund, "Sub-Saharan Africa's maternal death rate down 41 per cent," 2012. [Online]. Available: http://africa.unfpa.org/public/cache/offonce/news/pid/10767.

[15] World Health Organization, "Maternal Mortality," 2014. [Online]. Available: www.who.int/mediacentre/factsheets/fs348/en.

[16] S. Boseley, "Life expectancy around world shows dramatic rise, study finds," The Guardian, 13 December 2012.

[17] R. Freitas, "Nanomedicine, Natural Death, and the Quest for Accident-Limited Healthspans," 2004. [Online]. Available: http://www.nanomedicine.com/Papers/ImmInst2004.html.

[18] C. Hassen-Khodja, "Healthy Life Years," 2013. [Online]. Available: http://www.healthy-life-years.eu/.

[19] G. T. A. M. Milton Weinstein, "QALYs: The Basics," Value in Health, vol. 12, no. 1, pp. s5-s9, 2009.

[20] "EuroQol," 2015. [Online]. Available: http://www.euroqol. org/.

[21] M. Malek, "Implementing QALYs," March 2001. [Online]. Available: http://meds.queensu.ca/medicine/obgyn/pdf/what _is/ImplementQALYs.pdf.

[22] World Health Organization, "The Global Burden of Disease, 2004 Update," 2004. [Online]. Available: http://www.who.int/heal thinfo/global_burden_disease/GBD_report_2004update_full.pdf.

[23] P. Ehrlich and A. Ehrlich, The Population Explosion, Simon and Shuster, 1990.

[24] E. M. Lederer, "UN says half the world's population will live in urban areas by end of 2008," Associated Press, 26 Feb 2008.

[25] "Open-air computers," The Economist, 27 Oct 2012.

[26] The Peel Web, "A Web of English History," 2013. [Online]. Available: http://www.historyhome.co.uk/peel/social/popgra ph.htm.

[27] British Broadcasting Corporation, "GCSE Bitesize: Geography - The demographic transition model," 2014. [Online]. Available: http://www.bbc.co.uk/schools/gcsebitesize/geography/popul ation/population_change_structure_rev4.shtml.

[28] Dementia Research Group, "Demographic Aging," [Online]. Available: http://bit.ly/18peFL5.

[29] R. Corruccini and S. Kaul, "The epidemiological transition and the anthropology of minor chronic non-infectious diseases," Medical Anthropology, vol. 7, pp. 36-50, 1983.

[30] "Wikipedia," [Online]. Available: http://en.wikipedia.org/w iki/Epidemiological_transition. [Accessed 30 June 2013].

[31] A. Omran, "The Epdemiologic Transition: A Theory of the Epidemiology of Population Change," The Millibank Quarterly, vol. 83, no. 4, pp. 731-57, 2005.

[32] "World Hunger," [Online]. Available: www.worldhunger.org.

[33] World Hunger, 2015. [Online]. Available: http://www.worl
dhunger.org/.

[34] "International Food Policy Research Institute," 2014. [On-
line]. Available: http://www.ifpri.org/.

[35] International Food Policy Research Institute, "Renewed Policy Action for the Poorest and Hungry in South Asia," 2008. [Online].
Available: http://www.ifpri.org/sites/default/files/publications/cp1
0.pdf.

[36] "World Food Programme," 2010. [Online]. Available: http:/
/www.wfp.org/.

[37] P. L. Becker, "Measuring Commitment to Health: Global
Health Indicators Working Group Report," Center for Global Development, 2006.

[38] United Nations Department of Economic and Social affairs,
"World Population Prospects: The 2012 Revision," 2012. [Online].
Available: http://esa.un.org/wpp/unpp/panel_indicators.htm.

[39] The Wilson Center, "Yemen: Revisiting Demography After the Arab Spring," 2012. [Online]. Available: http://www.newsecuritybeat.org/2012/04/yemen-revisiting-d
emography-after-the-arab-spring/.

Chapter Twelve

ABOUT THE AUTHOR

R aywat Deonandan holds a PhD in Epidemiology & Biostatistics from the University of Western Ontario, and is a faculty member of the Interdisciplinary School of Health Sciences at the University of Ottawa. Since 2009, he has served on the Board of Directors for the Canadian Society of Epidemiology and Biostatistics. He is also the founder and Executive Editor of the Interdisciplinary Journal of Health Sciences.

www.ingramcontent.com/pod-product-compliance
Lightning Source LLC
Chambersburg PA
CBHW070946210326
41520CB00021B/7079